Japan Quake

Why do humans live in dangerous places?

Simon Saint

Revive Publications

*A catalogue record for this book is available
from the British Library*

ISBN: 978-1-907962-34-9

Published by Revive Publications

Reading, England

For a better future

Contents

Contents

Preface

I was stimulated to write this book by my desire to see a world in which there is much less human suffering than exists today.

Human suffering can be thought of as having two causes – other humans and 'nature'. In this book my concern is the attempt to understand and minimise the human suffering caused by 'nature'.

I will be considering two events which are typically thought of as 'natural disasters' that involved immense human suffering – the Boxing Day Tsunami and the current events in Japan – and I will be suggesting that these events have natural causes but that they are 'human-made' disasters.

This idea might surprise, even shock, you at first. But there is a benefit to seeing these human tragedies as 'human-made'; for, if they are 'human-made' then changes within the human-realm can stop them reoccurring in the future.

Introduction

For many years I have been thinking about the devastating effects that the workings of the planet sometimes have on the human species. When such devastating effects occur the event is typically referred to as a 'natural disaster'.

One event that particularly got me thinking was the Boxing Day Tsunami which occurred on 26 December 2008. This event was triggered by an earthquake off the coast of the island of Sumatra. The Tsunami which was triggered by the earthquake had devastating effects for the humans who live in the countries surrounding the Indian Ocean. Around

250,000 people are believed to have died due to this event.

In the aftermath of this tragedy it became clear that there were a number of reasons why so many people died. The event was triggered by the workings of the planet, but human actions were part of the reason that so many people died. Humans had destroyed large parts of the mangrove swamps which provided a layer of protection between the ocean and human habitation zones on the land. The areas where the mangrove swamps remained were significantly less damaged by the Tsunami than the large areas where human destruction had previously occurred.

Was the Boxing Day Tsunami a 'natural disaster' or a 'human-made disaster'? The answer is clearly that it was both. The initial cause was not caused by humans, but the effects on the human species were so bad, the death toll was so high, the damage to human homes was so extensive, because of human actions. Of this there is no doubt.

As I write these words another almost unbearably sad tragedy is currently affecting the human species. In contrast to the Boxing Day Tsunami (which is still widely referred to as a 'natural disaster') this tragedy is widely acknowledged to have both 'natural' and 'human-made' elements.

On Friday 11 March 2011 an earthquake off the coast of Japan triggered a series of tsunami waves which swept through a large area of Japan causing extensive damage to buildings and the infrastructure. As I write at least 10000 people are believed to have died due to this event. The human-made element of this event arises because the 'natural' event caused major damage to the nuclear power stations in Japan; in particular to the Fukushima I Nuclear Power Plant. A partial meltdown in some of the reactors resulted in a release of radiation from the plant. As I write several of the nuclear plant workers have died and a plethora of people are fleeing from Tokyo due to their fear of rising radiation levels.

These two events which I have outlined – the 2008 Boxing Day Tsunami and the current events in Japan – I am very saddened by. The suffering caused to humans by both events has been immense. I wonder whether anything can be done to prevent such immense suffering in the future. There are clearly human-made causes which make both events so devastating. My aim is to explore the extent to which humans can act so as to minimise the deleterious effects of the 'natural' workings of the Earth in the future.

The current events in Japan have caused me to think about the issue of the relationship between the human species and its habitation of the planet at a deeper level. Human actions preceding the Boxing Day Tsunami are clearly responsible for the scale of the disaster. Could it be that humans are actually far more responsible for *the devastation caused to humans* by 'natural' disasters than we usually admit?

In *Chapter One* I consider what an 'alien observer' would make of the Boxing Day Tsunami and the current events in Japan. Would it consider these events as 'natural disasters', 'human-made disasters', or a combination of the two? In *Chapter Two* I outline the 'hierarchy of danger' – the range of natural dangers that humans face from the workings of the planet. In *Chapter Three* I consider the question of why it is that humans live in dangerous places. Finally, in *Hope for the Future,* I end with a few words outlining why I have hope for the future.

Chapter 1

The Alien Observer

In order to get you to think about the relationship that exists between the human species and the planet I would like to get you to consider a 'thought experiment' which I am about to present to you. The purpose of this 'thought experiment' is to explore the extent to which 'natural disasters' are 'human-made' and the extent to which they are 'natural'.

Before I describe the 'thought experiment' I would like to clarify the terminology which is being used. I take the terms 'natural disaster' and 'human-made disaster' to be mutually exclusive. That is to say, a 'natural disaster' is a disaster that *is not* caused by human actions; contrarily, a 'human-made' disaster is a disaster which *is* caused by human actions. In effect, the 'natural' is that which is non-human.

What is a 'disaster'? A disaster is an event which involves great suffering (I will typically use this term to include death) to a great number of humans. So, if one human experiences immense suffering this is not a disaster. Whilst, if a small number of humans experience great suffering (for example, in a car crash) this is not a disaster. Furthermore, if a great number of humans suffer in a very moderate way then this also wouldn't be a disaster. So, as I am using the term 'disaster' (and I take it that this is the way the term is typically used) there has to be a great number of humans involved, and these humans have to suffer greatly.

So, to be clear, what this means is that earthquakes and tsunamis *themselves* are not 'disasters'. These natural workings of the planet are only 'disasters' if they lead to great suffering to a great number of humans. It is important to remember this.

The 'thought experiment' that I would like you to think about involves an alien observer. The alien observer is a super-intelligent life-form whose home is a distant galaxy. The alien observer has come to the Earth to observe the life-forms which exist on the planet in order to see how intelligent they are. You can, perhaps helpfully, think of the alien observer as observing the Earth from its spaceship which hovers above the Earth.

The alien observer has observed the Earth for a long time. Let us assume that it has observed the Earth since it was formed. The alien observer has an intimate knowledge of the workings of the planet. It has observed the evolution of life on the planet. It has observed the movements of the landmasses – from Pangaea (the 'super-continent') to the current locations of landmasses that we are familiar with today. The alien observer understands that the movements of the landmasses on the Earth are caused by plate tectonics – it observed such movements long before the human species came into existence. It has observed the evolution and demise of the dinosaurs, Ice Ages and Interglacial Periods, volcanic eruptions, mass extinctions, and the coming into existence and evolution of the human species.

What would the alien observer make of the human species? It has observed the human species evolve; it has observed the human species develop tools such as the wheel; it has observed the human species use animals for agriculture; it has observed the human species go through the scientific and industrial revolutions; it has observed the human species spread out over almost all of the landmasses of the Earth; it has observed the human species dividing the landmasses of the planet up into states; it has observed groups of humans killing other groups of humans *en masse;* it has observed humans acquiring knowledge about plate tectonics, volcanoes, sea-floor spreading, earthquakes, tsunamis, global warming and the depletion of the ozone layer. Well, I assume it would think a multitude of different things concerning the human species.

However, the point of the thought experiment is to focus our attention on what the alien observer would think about one particular thing. What would the alien observer think about the two events that I have outlined? What would the alien observer make of the Boxing Day Tsunami? And, what would the alien observer make of the current events in Japan – the earthquake and tsunami on Friday 11 March 2011, and their aftermath?

Let us first consider the Boxing Day Tsunami. The alien observer, let us remind ourselves, has seen a great number of earthquakes and tsunamis. These events are very regular occurrences in the Earth's history, and the human species are only very recent arrivals to the Earth. The alien observer will have observed the effects of a multitude of earthquakes and tsunamis. It will also have observed the land-masses of the Earth generating a layer of protection to guard against the worst effects of tsunamis – the mangrove swamps.

So, the alien observer will have observed tsunamis hitting the landmasses of the Earth before mangrove swamps existed. And, it will also have noticed that after the mangrove swamps came into existence the effects of the tsunamis on the land-masses were much reduced – the mangrove swamps were protecting the land behind them.

What would the alien observer make of two events? Firstly, humans building homes near the coast in areas where the alien observer knows that there are regular earthquakes of great magnitude and associated tsunamis? Secondly, humans then destroying the mangrove swamps which give protection to the land from tsunamis?

Well, when humans first started living near 'dangerous' coasts without knowing about the danger that they faced, the alien observer, we can assume, would feel very sorry for them – knowing that either these humans, or future humans (if humans continue to live there) will be hit by a tsunami. Furthermore, if humans destroyed the 'protecting' mangrove swamps without knowing about their protective role then, again, the alien observer would presumably feel very sorry for them, knowing that this will increase the devastation to humans when the tsunami hits.

Of course, the human species has acquired knowledge of plate tectonics, knowledge that earthquakes and tsunamis regularly affect certain landmasses, and knowledge that mangrove swamps provide protection from tsunamis. What will the alien observer make of the fact that despite having this knowledge humans still live in very dangerous places and that they make these places even more dangerous by removing the natural protection of the mangrove swamps? I presume that the alien observer will be rather bewildered – it knows that the human species is reasonably intelligent, that it has knowledge about plate tectonics and the protective role of mangrove swamps, yet humans destroy the swamps and still live near the coast in these dangerous places. It will surely be wondering why human beings live in such dangerous places. Do they like danger? Are they stupid? Do they like to suffer? The alien observer will be pondering these questions.

When it observed the Boxing Day Tsunami, and the disaster which followed – the immense number of human deaths and destruction to human homes – the alien observer will have no doubt that this disaster is human-made. The earthquake and tsunami are natural events – but the disaster arises because humans have decided to live in dangerous places and they have made these places even more dangerous than they were before (through removing the mangrove swamps)!

What would the alien observer make of the current events in Japan – the earthquake and the tsunami on Friday 11 March 2011, and their aftermath? The alien observer has seen many massive earthquakes and tsunamis hit this area, and *it knows that the human species knows* how dangerous a place this is to live – "Why do they still live there?" the alien observer asks itself. "Why don't the humans live in less dangerous places?" it wonders.

One can surely understand why the alien observer would conclude that the current events in Japan are a human-made disaster. If the human species lacked the knowledge of tectonic plate movements, and lacked the knowledge of which parts of the world are particularly susceptible to large earthquakes and tsunamis, then the alien observer would conclude that this event is a natural disaster. If the human species has no knowledge of the danger then they do not play a part in the disaster which materialises. However, if one has knowledge that cars are dangerous, that cars exist on busy roads, and one then walks into a busy road, then one clearly has knowledge that one is in a dangerous place; one has to consider it one's own (human) fault if one dies when one is hit by a car.

So, why is the alien observer so bewildered? Why do humans live in dangerous places?

Chapter 2

The Hierarchy of Danger

I would now like to introduce you to the 'hierarchy of danger'. There is nothing mysterious or complex about this term or the phenomena in the world that it refers to. It is just a useful term to highlight what is obvious.

There are clearly different types of danger. Danger can exist at a very low level, at a very high level, or somewhere between these two extremes. Danger can also exist in a very small part of the planet, across the entire Earth, or somewhere between these two extremes.

No parts of the Earth are danger free. An asteroid from outer space could collide with any part of the Earth and lead to a disaster in that part of the Earth. So, the Earth is a dangerous place to live. But to say this isn't to say very much. Everyone knows that life is a frail and precarious thing – one's life could end at any moment from any one of a plethora of dangers. However, the likelihood that one's life, or the life of any other human, will be severely affected (or ended) by an asteroid hitting the Earth is extremely low. The alien observer wouldn't be bewildered as to why humans live in dangerous places because some humans die when an asteroid strikes the Earth. Such an event would be purely a natural disaster.

However, it should be noted that, if human technology becomes advanced enough then humans can clearly prevent asteroids from hitting the Earth via their technology. If such a human capability exists, but is not used to prevent an incoming asteroid, then the resulting deaths obviously result from humans deciding not to use their technology. So, in this case, the disaster resulting from the asteroid strike would be a 'human-made' disaster. If such a capability existed, but was not used, then the alien observer *would* be bewildered. "Why have humans caused this disaster to happen?" the alien observer would be thinking. Do they like destruction? Are they stupid?

The alien observer fully understands that the human species needs to live on the Earth (it observed them evolve on the planet and thereby be suited to the environment of the planet), and it also understands that all life is precarious and faces a constant danger of death (from asteroids, other life-forms, disease, etc.). What it has trouble comprehending is why humans, despite having the knowledge that some parts of the Earth contain a low level of danger and other parts contain a high level of danger, still live in places with a high level of danger.

The alien observer itself evolved on a planet in a distant galaxy. Its species is the most intelligent form of life on its home planet. They initially inhabited all of the landmasses on their planet, but there was a devastating event in which an earthquake and tsunami on the planet caused the death of

200,000 'aliens'. After this event, they investigated their planet and came to understand about the operation of plate tectonics and the dangers that the workings of their planet presented to their continued existence. The alien species wanted to survive, and they didn't want members of their species to be wiped out in another disaster due to the tectonic movements of the plates of their planet. The suffering was too much to bear for a second time. So, they decided to live on parts of the planet where the danger was low, and not to live on parts of the planet where there was a high danger of disaster due to the workings of the planet. By taking this action there has been no alien deaths resulting from earthquakes and tsunamis for 87,000 years. Suffering has been reduced and the planet is a much more pleasant place to live.

One can understand the exasperation and bewilderment of the alien observer as he observes the Earth. "Are these humans not like us?" it thinks. We don't like to suffer. We don't like disasters. We don't like fellow members of our species to die *en masse.* "Why do these humans continue to live in places they know are dangerous?" it thinks. Do they like disasters? Do they like to suffer? Are they stupid?

Let us return to the hierarchy of fear. Imagine that the hierarchy is a pyramid. At the wide base of the pyramid are the plethora of low level natural risks (for suffering or death to humans) that exist across the entire planet. These include the potential for disease or illness, the potential for accident, the potential for being attacked by a non-human life-form, and the potential for being hit by an asteroid or a bolt of lightning. Wherever humans live on the Earth a great number of these natural dangers exist.

At the narrow top of the pyramid there are a very few extreme natural dangers. These are dangers which are not common – most landmasses of the Earth are not affected by these dangers. They are dangers which have specific locations which are determined by the workings and make-up of the landmasses of the Earth. There are parts of the Earth which are more dangerous to live because of human social factors – City A has a very high murder rate, while City B has an extremely low murder rate, so City A appears to be a more dangerous place to live. These factors might well play a part in determining where one wants to live but here I am solely concerned with exploring the 'natural' dangers that affect certain landmasses (a natural danger could affect City A but not City B).

So, at the top of the pyramid are places which present extreme dangers to those humans who live there due to natural factors. To take an obvious example, the edges of the tops of cliffs are very dangerous places. When I am going for a coastal walk I try to leave a small gap between the cliff edge and where I am walking. This is obviously because if I was walking right on the cliff edge then this would be much more dangerous. A slip, a momentary lapse of concentration, could see me toppling over the cliff edge and falling to my death.

Imagine what it would be like to live on the top of a cliff at the edge of the cliff. Imagine that one's house is right on the cliff edge and that in order to get out of one's house one has to use a door which is facing the cliff edge, and that there is only a narrow ledge which one can walk along to get to an 'inland' location. If there was a whole community of humans with such houses it would be inevitable that some humans would regularly fall over the cliff edge into the ocean (and thereby experience great suffering or death) as they were attempting to leave their homes. The ledge is narrow and slipping is going to occur occasionally as people lose their balance, particularly when caught by a gust of wind.

What would the alien observer make of such a community of cliff-edge dwelling humans? I take it that the alien observer would be rather bewildered? "Why don't they move inland to a less dangerous location so that they can leave their homes much more safely?" it thinks. "Do they like to suffer?" "Do they like to die?" "Are they stupid?"

There are, as far as I am aware, no such cliff-edge dwelling communities on the Earth. The point I am making is simply that some places are more dangerous to live than others due to the make-up and workings of the landmasses of the Earth. When we understand how the landmasses work we can decide not to live in the most dangerous places and thereby reduce the occurrences of suffering and death.

So, the places at the top of the pyramid include places such as the edges of cliffs. If one was to walk along such a cliff edge then one would face an extreme danger, and the more time one decided to continue this activity the greater will be the danger that one faces.

There are other places at the top of the pyramid which are not so obvious. One does not need a high level of knowledge concerning the workings of the planet to realise that walking along a cliff edge is an extremely dangerous thing to do! When one understands that the interior of the Earth contains a molten core which is constantly swirling around and that the landmasses that we live on are a thin crust which is sitting on top of this swirling activity then one can see a whole new range of dangers. The surface of the Earth is covered with a number of tectonic plates. These plates emerged out of the

37

swirling activity below and they are continuously returning to the swirling activity below; some plates lie on top of other plates and the lower plate gradually gets transformed and absorbed back into the swirling mass from which it first arose.

These movements of the plates are the cause of earthquakes, and earthquakes are the cause of tsunamis. The swirling activity from below also sometimes comes to the surface when volcanoes erupt. Living on an active volcano is obviously a much more dangerous place to live than most of the other landmasses of the Earth. Because the plates have edges where they collide, and we know where these edges are, we can work out which parts of the Earth are particularly susceptible to large earthquakes and tsunamis. That is to say, we can work out which parts of the Earth are at the top of the pyramid – which parts of the Earth are extremely dangerous places to live. When the species of which the alien observer is a part acquired such knowledge about their own planet they decided to stop living in such extremely dangerous areas.

The plethora of low level dangers which form the base of the hierarchy/pyramid of danger cannot be escaped. However, the human species clearly has the option of whether or not it faces the high level dangers which exist at the top of the hierarchy/pyramid of danger.

Chapter 3

Why do Humans Live in Dangerous Places?

Now that we have explored the hierarchy of danger we can return to our earlier question of why the alien observer is so bewildered. The alien observer asked itself: "Why do humans live in dangerous places?" Now that we have seen that everywhere is dangerous we can understand that what the alien observer is bewildered by is more accurately phrased as the following question: "Why do humans live in *extremely* dangerous places?

To try and understand why this is so a brief consideration of human history is necessary. As far as we can tell the human species evolved in Africa and then started to explore the parts of the Earth nearest to Africa (the only places it could get to by walking!). As time passed members of the human species continued to explore and they found and populated an increasing amount of the landmasses of the Earth. When boats were first invented then this enabled humans to find and populate secluded landmasses located 'deep' within the oceans.

The motivation for finding and populating all possible landmasses on the Earth seems to be a simple one. Firstly, we are an intuitively curious species and like to 'find everything that is out there' both in the wider Earth and beyond. Secondly, because if one was part of a group of humans that first discovered a landmass then one could stake a claim to it. Not only could one use all of the resources which existed in this area of the Earth, but one could attempt to 'own' the land. One could settle down and populate the landmass – claim a stake to 'ownership' of the land. If someone tried to take the land and resources for themselves then fighting is likely to have arisen. Whichever side lost (if they still had any members left!) then had a motivation to find an alternative landmass which they could 'make their own' and attempt to defend from other groups of humans.

So, I am suggesting that there was a strong motivation for humans to populate and 'own' every possible area of land on the Earth. When all of this occurred there was obviously knowledge of some extreme dangers (dangers at the top of the pyramid/hierarchy) – dangers such as the risk of walking along the edge of a cliff. However, there wasn't knowledge concerning the movements of the plates and the causes of earthquakes and tsunamis. It was simply the case that more or less any land-mass was considered to be a suitable place to populate, to 'own' and to defend.

Relatively recently this 'ownership' of the landmasses became formalised through the creation of nation states with fixed boundaries. In this way the entire landmass of the Earth has become carved up with particular people entitled to live on a particular piece of land, but typically unable to live elsewhere (you might be able to leave the landmass for a couple of weeks for a holiday if you have a passport, and/or you might be lucky enough to be able to relocate and subsequently reside in an alternative landmass).

With the creation of nation states people are generally 'stuck' to a particular landmass. This might have seemed to be a good idea at the time. However, now we have acquired the knowledge that some of these landmasses are very dangerous places to live. Merely being on some landmasses means that one is at the top of the hierarchy/pyramid of danger. Our knowledge of plate tectonics means that we can tell that some landmasses, due to their location *vis-à-vis* the plates, are going to experience very severe earthquakes and tsunamis.

We know this, but the people who happen to live in these areas supposedly cannot live somewhere less dangerous. This is their nation state –

they cannot leave. They have just 'got the short straw' and must live with the danger, suffering and destruction that this entails. This seems to be where we are at present.

Of course, it doesn't have to be this way. The alien observer is bewildered because its species does not have states – they are not forced to live in extremely dangerous places. When they realised that certain parts of their planet were extremely dangerous places to live they simply moved to less dangerous landmasses.

Could the same thing happen on the Earth? Of course it could. There is no reason why, if we wanted, that all humans could live in places where there is only a low level of danger from the workings of the planet. This would seemingly require a redrawing of the map of states, and possibly the forging of states together (a new state being created by the 'incorporation' of a high-danger state into a low-danger state). Although in many cases changes within a state would be sufficient. Members of a state could simply decide that a particular part of their landmass should not be inhabited due to the extreme danger, while the rest is habitable.

If this were to occur the alien observer would no longer be bewildered. The earthquakes and tsunamis would still occur but there would be no more (or at least far fewer and less severe) disasters because the landmasses affected would be uninhabited. If this were to occur the suffering and misery within the human realm would be greatly reduced. Surely this would be a good thing?

Presumably the human species wants to reduce the number of disasters resulting from the workings of the Earth (recall that the *workings* are natural, but that the *disaster* is 'human-made' when the knowledge of the extreme danger exists, but no action is taken to stop it causing great human suffering/death). Surely less human suffering is a good thing?

We currently seem to be living in a strange state of transition. Firstly, we inhabited all possible landmasses. Secondly, we became 'stuck' to these landmasses. Thirdly, we realised that some of these landmasses are extremely dangerous places to live. Fourthly, we don't want (ourselves or our fellow humans) to live in extremely dangerous places. Fifthly ...well the fifthly hasn't occurred yet.

The 'fifthly' will, presumably, occur in the future when we follow the example of the species of the alien observer. In this stage we decide to use our knowledge to reduce (ideally eliminate) the 'human-made disasters' which occur due to the natural workings of the planet.

When we have the knowledge about the workings of the planet – the knowledge that certain parts of the Earth are extremely dangerous places to live – it is inexcusable to continue to refer to the immense suffering and human deaths which follow as the result of a '*natural* disaster'.

The workings of the planet are natural. There is nothing natural about immense human suffering and death occurring because humans decide to live in extremely dangerous places.

Let us make the alien observer bask in our intelligence. Let us not bewilder it any longer. There is no reason why the alien should be thinking: "Do they like to suffer?" "Do they like to die?" "Are they stupid?" Hopefully, in the near future, the alien observer will be thinking: "They have learned... they don't like to suffer... they are intelligent after all."

Hope for the Future

It is easy to envision a future in which humans suffer much less from the workings of the planet than they do today.

Let us hope that this future arrives soon.

Perhaps the great suffering which occurred in the aftermath of the Boxing Day Tsunami, and the great suffering which is currently ongoing in Japan, will help to catalyse the changes which are required.

Other books by the author:

Reflections on Humans and their Surroundings
(2011)

Living a Spiritual Life: Moving to a higher-self
(2011)

117 Tips for a Spiritual Life
(2011)

www.ingramcontent.com/pod-product-compliance
Lightning Source LLC
Chambersburg PA
CBHW071343290326
41933CB00040B/2156